flavouring with
Olive Oil

flavouring with

Olive Oil

Clare Gordon-Smith

photography by

James Merrell

RYLAND
PETERS
& SMALL

Art Director **Jacqui Small**

Art Editor **Penny Stock**

Design Assistant **Mark Latter**

Editor **Elsa Petersen-Schepelern**

Photography **James Merrell**

Food Stylist **Clare Gordon-Smith**

Stylist **Sue Skeen**

Production Consultant **Vincent Smith**

**Our thanks to Christine Walsh and Ian Bartlett,
Jenny Merrell, Peter Bray and Robert Roseman.**

First Published in Great Britain in 1996
Reprinted 1997
by Ryland Peters & Small
Cavendish House, 51-55 Mortimer Street, London W1N 7TD

Text © Clare Gordon-Smith 1996
Design and photographs © Ryland Peters & Small 1996
Produced by Toppan Printing Company
Printed and bound in Hong Kong

ISBN 1 900518 00 7

A CIP record for this book is available from the British Library

Notes:
**Metric and imperial measurements are
given. Use one set of measurements
only and not a mixture of both.**

**Ovens should be preheated to the
specified temperature – if using a
fan-assisted oven, adjust time and
temperature according to the
manufacturer's instructions.**

Olive oil is one of those rare commodities – something that tastes wonderful and is actually good for you. Specialist food shops conduct olive oil tastings, like wine tastings. The oils are poured into small white dishes and accompanied by broken bread. The oils are tasted by dipping the bread into each oil and comparing the flavours. Oils at such **tastings** are offered labelled according to their country or region of origin, and also to their grade. The grades of olive oil are determined by the method of harvesting and pressing. The most desirable in terms of quality are hand-harvested, first pressed and cold pressed. The finest oils from each region are harvested and pressed in this way.

The best oil is **extra-virgin**, with an acidity of less than one per cent. It is best used as a flavouring agent – in dressings, or poured over finished dishes, so its inherent taste is not altered by heat. Other grades include fine virgin olive oil, semi-fine olive oil, refined and pure olive oil, and you should choose one of these lower grade oils for cooking and for making mayonnaise. **Aïoli**, the rich garlic mayonnaise of Provence, should however be made from good green extra-virgin oil.

the flavours of
Olive Oil

Olives are grown all around the Mediterranean, as well as in South Africa, Australia and California. The top oil producers are Spain, Italy, Greece and France. Each country's oil has a distinctive **flavour**, with nuances and colour that vary between regions – affected by climate, soil and fruit

variety. French oil is very **sweet** and, though produced in small quantities, is of fine quality. Greek oil is more **aromatic**, and oils from the Pelopponese and Crete are very highly regarded. Spain is the largest producer of oil and its oil tastes very **fruity**.

8 The flavours of olive oil

Fine Italian oils are produced
in Liguria, Tuscany, Umbria and Apulia.
Tuscan oils have a **peppery** aftertaste, which
varies from district to district.
Californian olive oil is fruity and **mild**, and
produced under varietal names such as
Manzanilla, Sevillano and Mission.

Australian oils are labelled, like their **wines**,
with fruit varieties, climate, aspect of the
groves and pedigrees of the trees.
Cyprus produces some superb oils, while oil
produced in Israel, the Lebanon, Turkey, Syria,
Tunisia, Portugal, Morocco, Algeria and South
Africa is mostly consumed in the home market.

The flavours of olive oil 9

Starters

Chilled tomato soup
with basil pesto

A soup for high summer when tomatoes are
at their best. Serve with garlic croûtons –
don't bother to dice them neatly – just tear
the bread roughly into pieces, fry with garlic
in olive oil until lightly golden, then drain.

Place the tomatoes in a blender or food processor
and purée into a mush.

Pass through a stainless steel sieve into a bowl,
pushing through as much juice and pulp as possible,
giving a creamy consistency.

Thin if necessary with the olive oil and tomato juice.
Mix in all the remaining ingredients, place the bowl
inside a larger bowl, and fill the space between the
two bowls with ice. Pour cold water over the ice
and chill for about 2 hours to develop the flavours.
Serve in soup bowls, garnished with the basil pesto,
strips of roasted yellow peppers, a drizzle of olive oil
and an ice cube, together with a bowl
of garlic croûtons.

1 kg/2 lb very ripe
plum tomatoes

3 tablespoons olive oil

150 ml/¼ pint
tomato juice

7 cm/3 inch
piece of cucumber,
finely chopped

1 yellow pepper,
roasted, skinned,
deseeded and chopped

1 small red onion,
finely chopped

1 tablespoon
balsamic vinegar

2 tablespoons torn
fresh basil leaves

salt and freshly
ground black pepper

to serve

basil pesto
(see page 21)

roasted yellow peppers

extra-virgin olive oil

ice cubes

garlic croûtons

Serves 4

Tuscan bean soup
with green pesto

A rustic soup, with a delicious green pesto swirled through just before serving. If broad beans are out of season, substitute green flageolet beans or more white cannellini beans. If you're short of time, you could also use canned cannellinis or flageolets.

To prepare the dried beans, soak overnight in cold water. Next day, drain, place in a large saucepan with cold water to cover, bring to the boil and simmer gently until the beans are tender (about 30 minutes, depending on the age of the beans). Add a little salt about 5 minutes before the beans are done. To make the soup, heat the olive oil in a large, heavy-based saucepan, add the onion and cook over a medium heat until softened and transparent. Add the cabbage and fry for a few minutes. Add the stock, bring to the boil and simmer, covered, for about 20 minutes. Add the broad beans and courgettes and simmer for a further 10 minutes. Stir in the oregano and season with salt and freshly ground black pepper.

Add the cannellini beans, with their cooking liquid, and simmer together for about 5 minutes. To serve, divide the soup between 4 large, heated soup bowls, place a tablespoon of pesto on top, together with a sprig of basil, if using, and scatter with the roughly crumbled Parmesan.

125 g/4 oz dried white cannellini beans

150 ml/¼ pint pure olive oil

1 onion, finely chopped

¼ cabbage, finely sliced

1 litre/1¾ pints vegetable stock

250 g/8 oz broad beans (about 500 g/1 lb in the pod)

2 courgettes, sliced and quartered

2 tablespoons snipped fresh oregano leaves

salt and freshly ground black pepper, to taste

to serve

4 tablespoons basil pesto (see page 21)

4 sprigs of basil (optional)

4 tablespoons crumbled fresh Parmesan cheese

Serves 4

a **rustic** italian soup, with brilliant green basil pesto

Goats' cheese
and Bordeaux lentil salad

Bordeaux lentils have a robust taste and stay whole and firm when cooked – their flavour contrasts well with the strong goats' cheese. Puy lentils, darker in colour, are more widely available in good quality food stores.

Place the lentils in a saucepan and cover with cold water. Add the garlic and half the parsley, bring to the boil and simmer for 10 minutes.

Add the onion, celery and a pinch of salt, and simmer for a further 10–15 minutes, adding a little hot water if required, until the lentils are just cooked but still firm. Discard the herbs and garlic and toss the lentils in olive oil while still warm.

Crumble the goats' cheese and mix into the lentils. Add the snipped chives, the lemon juice, the remaining parsley, chopped, with salt and freshly ground black pepper to taste.

To serve, place salad leaves on each plate, spoon over the lentil and cheese mixture and garnish with cherry tomatoes, roughly chopped into chunks.

125 g/4 oz Bordeaux
or Puy lentils

1–3 garlic cloves

1 bunch of
flat leaf parsley

1 red onion, finely diced

1 celery stalk,
finely diced

pinch of salt

250 ml/½ pint
virgin olive oil

250 g/ 8 oz
goats' cheese

1 bunch of chives,
snipped

50 ml/2 fl oz lemon juice

salt and freshly
ground black pepper

to serve

salad leaves, such
as rocket, spinach,
watercress or
lambs' lettuce

250 g/8 oz
cherry tomatoes

Serves 4

a very **eas**

with truffle oil to give th

Leek and new potato salad
with truffle oil

Truffle oil is the most important ingredient in this recipe, and is also delicious drizzled over pasta. Buy it in small quantities, such as the little bottles of white and black truffle oils pictured, because they will last for ages. Miniature baguette leeks always look pretty in salads, but if you can't find them, use larger leeks cut into chunks and leave the chunks intact when serving.

Cook the potatoes in boiling salted water for about 20 minutes, until tender.
Trim the leeks and slice into 2.5 cm/1 inch chunks, if large. Wash thoroughly and cook in boiling salted water for 5–7 minutes, until just tender.
Place the parsley, olive oil, salt and pepper in a bowl. Drain the potatoes, slice thickly and add to the dressing while still warm, so they absorb all the flavours.
Drain the leeks thoroughly and add to the potatoes. Stir well, taste and adjust the seasoning.
To serve, spoon the salad on to serving plates and drizzle with truffle oil.

750 g/1½ lb new potatoes

500 g/1 lb young leeks (preferably baguettes)

2–4 tablespoons roughly chopped fresh flat leaf parsley

3 tablespoons olive oil

sea salt and freshly ground black pepper

4 tablespoons truffle oil, to serve

Serves 4

ut spectacular salad—

xtra dash of luxury

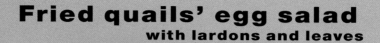

Fried quails' egg salad
with lardons and leaves

Eggs fried in olive oil have crispy bottoms
and lacy edges, but you can also use boiled
eggs. Nasturtium flowers make a colourful,
peppery addition to this salad in summer.

mixed leaf salad

4 tablespoons olive oil

8 quails' eggs

**4 slices streaky bacon,
cut into strips**

**2 tablespoons
red wine vinegar**

**1 teaspoon
wholegrain mustard**

**salt and freshly
ground black pepper**

Serves 4

Arrange the leaves on 4 starter plates.
Heat 1 tablespoon of the oil in a small frying pan
and lightly fry 4 of the eggs. Carefully place on the
leaves. Add another tablespoon of oil and repeat
with the remaining eggs. Add the bacon pieces to
the pan, fry until crisp, then add to the salad.
Whisk in the vinegar, mustard and the remaining oil,
and pour over the eggs and leaves. Season with salt
and freshly ground black pepper.
Serve with crusty bread.

Bruschetta
with tomatoes and olives

A perfect fast snack – like a good-looking
pizza – and also delicious as a first course
for a summer lunch. Other good toppings are
the goats' cheese on page 29, the artichokes
on page 26 or the aubergines on page 58.

Split the focaccias in half and toast lightly.
Spread with a dollop of green basil pesto, then
scatter over the tomatoes and olives. Add a slice of
mozzarella to each, then drizzle with olive oil and
place under the grill for a few minutes.
Serve, garnished with sprigs of fresh basil leaves.

4 focaccias or sliced
baguette loaf

green basil pesto
(see opposite)

175 g/6 oz vine-ripened
cherry tomatoes

50 g/2 oz black olives,
pitted and quartered

2–4 mozzarella
cheeses, sliced

Italian single estate
olive oil

sprigs of basil,
to garnish

Makes 4

Olive tapenade

If available, use olives and olive oil from the
same region or country of origin, to intensify
the fruity flavour of the oil.

Pit and chop the olives, then place in a food
processor with the capers, anchovies and garlic, and
blend thoroughly. Add the olive oil gradually until the
mixture reaches a spreadable consistency.
Slice and toast the baguette, spread with the
tapenade and serve with grilled tomato slices.

175 g/6 oz
black olives

2 tablespoons capers

4 anchovy fillets,
rinsed and chopped

2 garlic cloves, crushed

150 ml/¼ pint olive oil

to serve

1 baguette loaf

grilled tomato slices

Serves 4

Green basil pesto

Although ready-made pesto is widely available, it's never as good as the home-made variety. Whenever you see some good fresh basil, or if you grow your own, try this recipe – it's a revelation!

Place all the ingredients in a blender or food processor and purée to a thick cream, adding a little more oil if it is too thick. Add salt to taste. Spoon into a jar, and float a thin layer of oil on the top. Store in the refrigerator until ready to use.

125 g/4 oz fresh basil leaves

25 g/1 oz pine nuts

25 g/1 oz fresh Parmesan cheese, grated

150 ml/¼ pint olive oil

salt

Makes 1 jar of 300 ml/½ pint

Flavoured oil

You can buy flavoured oils, but it's a simple matter to make your own. Make them in very small quantities, store in the refrigerator and use them quickly. Use any pretty, stoppered bottles, and sterilize both bottles and corks thoroughly before using. It is best not to make garlic or truffle oils yourself.

If using herbs, wash them thoroughly before use. Pour the oil into a saucepan and gently heat together with the herbs or flavourings. Pour the hot oil into sterilized bottles, then add the heated herbs or flavourings. Store in the refrigerator and use within 24 hours.

virgin olive oil

flavourings

a choice of:
1 sprig of rosemary,
2 sprigs of thyme,
1 tablespoon black peppercorns,
2 sprigs bay leaves,
2 pieces of lemon peel,
or 1 tablespoon dried chillies

Left, front row, from left, Bruschettas with marinated artichokes (recipe page 26), with olive tapenade (page 20) and with green basil pesto (page 21). Right, with green basil pesto, and with tomatoes and olives (recipe page 20 – see also back jacket).

Char-grilled asparagus
with a warm bread salad

The Italians use bread soaked in oil or vinegar as a base for sauces and soups. Char-grilling or roasting asparagus keeps the spears crisp on the inside. Caperberries are delicious if you can find them – but ordinary capers also work well.

To make the salad, tear the bread into chunks and mix with half the vinegar until softened. Squeeze dry and mix with the egg, parsley, caperberries or capers, salt and pepper. Heat the oil and remaining wine vinegar in a pan until warm, then stir in the bread mixture. Set aside to keep warm while you cook the asparagus.

Heat a grill pan on top of the stove and lightly brush with olive oil. Add the asparagus spears and cook on all sides until slightly shrivelled and brown. Alternatively, place in a roasting tin and cook in a preheated oven at 200°C (400°F) Gas Mark 6 for about 5–7 minutes until browned.

Place on a warmed serving plate, spoon over the warm bread salad, sprinkle with the balsamic vinegar, and serve immediately.

500 g/1 lb asparagus

oil, for brushing

2 teaspoons balsamic vinegar, to serve

warm bread salad

75 g/3 oz good bread (sourdough or country style)

125 ml/4 fl oz white wine vinegar

1 hard-boiled egg, quartered

1 bunch of flat leaf parsley, roughly chopped

50 g/2 oz caperberries, (if available), or capers packed in salt

150 ml/¼ pint virgin olive oil

salt and freshly ground black pepper

Serves 4

Baby artichokes
marinated in olive oil and lemon

These wonderfully flavoured artichokes
can be made in advance and stored in
sealed jars. Serve them in this salad,
or alternatively as a bruschetta topping,
or as an antipasto with sliced salami.

Peel away the outer layer of the artichoke stems and
trim the leaves with a knife or scissors. (Young,
tender artichokes have no prickly chokes.)
Brush all the cut surfaces with a little lemon juice
to prevent browning.
Place in a shallow pan with a tightly fitting lid, add
the lemon juice, bay leaves, peppercorns, parsley,
olive oil, sea salt and just enough water to cover.
Bring to the boil and simmer for 10–15 minutes until
tender. Marinate in the refrigerator for at least
24 hours. When ready to serve, drain and reserve
the liquid to use again as a vinaigrette.
Cut the artichokes in half, heap the spinach or
mustard leaves on the toast, place the artichoke
halves on top and pour over some of the marinade.

1 kg/2 lb
baby artichokes

juice of 3 lemons

3 bay leaves

10 black peppercorns

1 bunch of
flat leaf parsley,
chopped

150 ml/¼ pint olive oil

sea salt

to serve

250 g/8 oz baby
spinach or
mustard leaves

4 slices sourdough
bread, toasted

Serves 4

serve as an **antipasto**, on bruschetta,

or as a salad with crusty bread

Goats' cheeses
marinated in olive oil

Marinate your own goats' cheeses in olive oil – they can then be used grilled on toast, as a pasta sauce with snipped herbs, or served with a selection of crisp wafers.

Place the whole goats' cheeses in a clean Kilner jar or stoneware pot. Pour in the olive oil, add salt, then push in the olives, if using. Blanch the garlic, chilli, herbs, spices and lemon peel in boiling water for 1 minute, drain then add to the jar. Cover and leave to marinate in the refrigerator for 2 days before using within 1 week.

500 g/1 lb small goats' cheeses, or large ones, sliced and quartered

600 ml/1 pint virgin olive oil

75 g/3 oz pitted black olives (optional)

4 garlic cloves

a selection of:
1 small red chilli ,
2–3 sprigs of fresh thyme or rosemary,
2 bay leaves and
1 tablespoon peppercorns

lemon peel

sea salt

Makes 2 jars of 300 ml/½ pint

Mushrooms and anchovies
with pan-fried polenta wedges

great for lunch

Polenta can be bought ready-made, or
prepared according to the recipe below.
As an alternative, use slices of focaccia,
drizzled with olive oil and baked in the oven
until crispy. You can also make this recipe
without the anchovies.

To make the polenta, place in a pan, add the boiling
water and prepare according to the packet
instructions. When done, spread into a deep pan to
set (about 5 minutes), then cut into wedges and fry
in the olive oil until lightly browned. Alternatively,
place the pieces on a baking tray, drizzle with a little
olive oil and cook in a preheated oven at 200°C
(400°F) Gas Mark 6 for about 10 minutes.
Rinse the porcini, soak in warm water for 20 minutes,
then drain on paper towels and roughly chop.
Reserve the soaking liquid.
Heat the oil in a frying pan, and sauté the garlic for a
few minutes until softened and transparent. Add the
porcini and sauté for a few more minutes. Add the
sherry, chopped anchovies and porcini soaking
liquid, bring to the boil, and simmer for about
5 minutes. Taste and adjust the seasoning.
Remove the mushrooms, then bring the sauce to
the boil and reduce to about 4–6 tablespoons.
To serve, spoon the mixture on to the warm polenta,
pour over the sauce and sprinkle with oregano.

as a starter – easy to make

and absolutely **packed with flavour**

8 polenta pieces or
home-made polenta
(see below)

about 125 g/4 oz dried
porcini mushrooms

3 tablespoons
virgin olive oil

1 garlic clove, crushed

1 tablespoon dry sherry

50 g/2 oz anchovies
in olive oil, finely
chopped

sea salt and freshly
ground black pepper

chopped fresh oregano
leaves, to serve

fried polenta

125 g/4 oz
pre-cooked polenta

250 ml/8 fl oz
boiling water

olive oil, for frying

Serves 4

Main courses

Char-grilled tuna
with red onion salad and gremolata

Marinating is a way of giving the fish extra flavour. This recipe is perfect for cooking on a barbecue in the summer, but you can get a similar effect using a grill pan on top of the stove. This recipe is also good with other big fish such as swordfish.

To make the gremolata, mix the parsley with the olive oil, lemon juice and zest, salt and freshly ground black pepper. Set aside.

To make the salad, place the sliced onions in a shallow dish, pour over the olive oil and balsamic vinegar and sprinkle with the herbs and seasonings. Set aside.

Brush the tuna steaks with the olive oil. Heat a grill pan on top of the stove and, when very hot, quickly sear the tuna on both sides, then continue to cook for about 5 minutes.

Divide the onion salad between 4 dinner plates with the tuna beside. Heap a spoonful of gremolata on each steak, and serve with wedges of lemon.

4 tuna steaks

150 ml/¼ pint olive oil

salt and freshly
ground black pepper

lemon wedges, to serve

gremolata

1 large bunch of
flat leaf parsley,
roughly chopped

125 ml/4 fl oz olive oil

juice and finely sliced or
grated zest of 1 lemon

salt and freshly
ground black pepper

red onion salad

3 red onions, sliced

2 tablespoons
extra-virgin olive oil

1 tablespoon
balsamic vinegar

3 tablespoons
finely chopped
fresh flat leaf parsley

salt and freshly
ground black pepper

Serves 4

Roasted cod
with olive tapenade

The strong flavour of this tapenade works
very well with cod or other fish that hold
their shape, such as haddock fillets.

4 cod steaks,
about 125 g/4 oz each

125 g/4 oz pitted
green olives

1 tablespoon capers
packed in salt

1 bunch of
flat leaf parsley,
chopped

8 anchovy fillets

3 tablespoons olive oil

sea salt and freshly
ground black pepper

steamed couscous,
to serve

Serves 4

Place the cod steaks in a roasting tin.
Mix the olives with the capers and parsley.
Spread this mixture on top of the cod steaks,
place 2 anchovy fillets in a criss-cross on top of each
steak, pour over the olive oil and season with salt
and freshly ground black pepper.
Cook in a preheated oven at 200°C (400°F)
Gas Mark 6 for 15–20 minutes until the fish
turns white and opaque.
Serve with steamed couscous.

Seared beef fillet
with olives, tomatoes and rocket

A spectacularly easy recipe, packed with flavour and goodness. The beef can be cooked in a grill pan on top of the stove, or very quickly in a searingly hot oven.

Cut slits into the surface of the beef and insert the garlic slices. Pour over the olive oil and sprinkle with freshly ground black pepper. Tuck the bay leaves underneath and set aside for a few minutes.

To prepare the rocket salad, place the leaves in a bowl, add the tomatoes and scatter over the olives, if using. Drizzle with the extra-virgin olive oil.

Heat a grill pan on top of the stove until very hot. Add the beef fillet and bay leaves and cook the beef for about 5 minutes on each side, depending on how rare you like your beef. If roasting the beef, cook in a preheated oven at 200°C (400°F) Gas Mark 6 for 10 minutes for rare beef, 15 minutes for medium, and 20 minutes for well done.

Discard the bay leaves and set the meat aside, covered, for 5 minutes, to allow the meat to rest, then cut into thick slices.

Place a pile of rocket leaves on each plate, top with the beef slices, add the sliced tomatoes and black olives, if using, then scatter with torn parsley and sea salt, and dress with a little more extra-virgin olive oil.

500 g/1 lb beef fillet

4 garlic cloves, sliced lengthways

150 ml/¼ pint olive oil

4 bay leaves

sea salt and freshly ground black pepper

to serve

1 large bunch of fresh rocket leaves

4 ripe plum tomatoes, sliced

handful of black olives, preferably *herbes de provence* (optional)

1 tablespoon torn fresh flat leaf parsley

extra-virgin olive oil, preferably from Provence

Serves 4–6

Provençal lamb daube

This is a quick and easy update of one of the
great classics from the South of France. Use
oil from Provence, if you can find it, for a
truly authentic flavour.

Place the lamb in a shallow dish.
Mix the marinade ingredients together and pour over
the meat. Place in the refrigerator overnight or up to
2 days, depending on time available. Remove from
the refrigerator 1 hour before cooking to return
to room temperature.
Remove the lamb from the marinade and pat dry with
kitchen paper. Reserve the marinade.
Heat the olive oil in a large frying pan and brown the
meat on all sides.
Transfer to a casserole, sprinkle the flour over the oil
and juices in the pan, stir and cook for about
1 minute, then add the brandy.
Bring to the boil, remove from the heat and pour
over the meat in the casserole dish.
Add the marinade and stock, adding water to cover
the meat if necessary, season well, and bring up to
boiling point on top of the stove.
Transfer to a preheated oven and cook slowly at
170°C (325°F) Gas Mark 3 for 1–1½ hours.
Serve with stewed flageolet beans and mashed
potatoes made with olive oil.

6 large lamb shanks,
left whole,
or 1 leg of lamb,
about 1.5 kg/3 lb,
cut into thick slices

3 tablespoons olive oil

2 tablespoons flour

125 ml/4 fl oz brandy

300 ml/½ pint vegetable
or lamb stock

salt and freshly
ground black pepper

marinade

4 garlic cloves

2 carrots

300 ml/½ pint red wine

2 sprigs of thyme

2 sprigs of parsley

2 strips of orange peel

4 tablespoons olive oil

Serves 6

Marinated chicken
with avocado and spinach salad

It is important to use free-range chicken to give a good flavour in such a simple recipe – or in any other! Serve it either hot or cold, with a salad such as this one.

Place the chicken in a single layer in a roasting tin, mix the remaining ingredients together and pour over the chicken. Marinate in the refrigerator for at least 30 minutes, or longer if possible.
Place in a preheated oven and cook at 200°C (400°F) Gas Mark 6 for 20 minutes, or until the juices run clear when the chicken is pierced with a skewer through the thickest part.
To make the salad, blanch the beans in boiling water and refresh in iced water. Using a teaspoon, scoop the avocado flesh into balls and place in a bowl with the beans, lettuce and spinach leaves. Whisk the oil with the vinegar, salt and freshly ground black pepper, pour over the salad and serve.
Serve the chicken with new potatoes, accompanied by the avocado salad.

6 pieces of
free-range chicken

juice of 1 lemon

125 ml/4 fl oz
virgin olive oil

75 g/3 oz pitted
green olives

1 tablespoon capers,
packed in salt

1 bay leaf

4 sprigs of rosemary

**avocado and
spinach salad**

125 g/4 oz green beans

1 ripe avocado

1 batavia lettuce or other
soft lettuce

125 g/4 oz baby spinach
leaves

150 ml/¼ pint
extra-virgin olive oil

50 ml/2 fl oz
balsamic vinegar

salt and freshly
ground black pepper

Serves 4–6

an easy dish for a midweek dinner party

Chicken breasts
with pancetta and rosemary oil

Rosemary oil is a wonderful flavouring ingredient. Use it for frying or roasting – or as a quick dressing for salads and steamed vegetables. Serve this dish with whole new potatoes, tossed in olive oil and crushed garlic, sprinkled with sea salt, then baked in the oven at the same time as the chicken.

4 free-range, skinless chicken breasts

4 slices acorn-fed Westphalian ham, or other high-quality ham

rosemary oil

125 ml/4 fl oz extra-virgin olive oil

6 sprigs of rosemary

3 garlic cloves

pancetta filling

125 g/4 oz pancetta, finely chopped

125 g/4 oz ricotta cheese

sea salt and freshly ground black pepper

Serves 4

To make the rosemary oil, place all the ingredients in a small saucepan, heat gently until tepid, then set aside for about 15 minutes until ready to use. To make the filling, mix the pancetta with the ricotta, salt and freshly ground black pepper.

Using a small, sharp knife, cut pockets down the sides of the chicken breasts and loosely pack with the filling. Wrap each breast in a slice of ham. Place the stuffed chicken in a roasting tin, pour over the rosemary oil and cook in a preheated oven at 400°F (200°C) Gas Mark 6 for 20 minutes. Serve immediately with oven-baked garlic potatoes and a green leafy salad.

Braised pheasant
with apples, fennel and sage

A winter dish which also works well with
chicken, guinea fowl, quail, or other birds.
Omit the fennel if it is unavailable – the dish
is equally good without it. You could also
leave out the flour – the golden sauce will be
thinner, but it will still taste terrific.

Heat the olive oil in a large frying pan, add the
pheasant pieces and cook until golden brown on all
sides. Transfer to an ovenproof casserole.
Add the bacon to the pan, fry until golden, then add
to the casserole. Add the flour, if using, to the pan,
stir well and cook for 1 minute until lightly browned.
Stir in the apple juice, stock, sage and parsley, bring
to the boil, then pour over the pheasant.
Add seasoning, stir in the fennel, and bring to the
boil on top of the stove. Transfer to a preheated oven
and cook at 200°C (400°F) Gas Mark 6 for about
45 minutes, or until just tender. Remove the herbs.
Melt the butter in a small pan and, when sizzling,
add the apple slices and a pinch of sugar, and fry
until golden. Stir into the cooked pheasant
just before serving.
Serve with a glass of warming red wine and
mashed potato made with garlic.

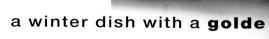

a winter dish with a golde

about 150 ml/¼ pint
olive oil

2 pheasants, cut into 8

6 slices streaky bacon,
cut into small pieces

1 tablespoon flour
(optional)

150 ml/¼ pint
apple juice

300 ml/½ pint
chicken stock

2–3 large sprigs of sage

6 large sprigs of
flat leaf parsley

1 fennel bulb,
thickly sliced

25 g/1 oz butter

2 Cox's apples, sliced

pinch of sugar

salt and freshly
ground black pepper

Serves 4–6

auce that tastes terrific

Pasta

Pappardelle
with olive and parsley sauce

A good-quality dried egg pasta is best for this recipe, though fresh pappardelle from a very good Italian deli could also be used.

To make the sauce, heat 1 tablespoon of the olive oil in a pan, add the shallot and fry until softened and transparent. Add the remaining sauce ingredients, simmer for 10 minutes, then place in a blender or food processor and purée until smooth.
Bring a large pan of salted water to the boil, add the pasta and cook according to the packet instructions until just *al dente*.
Drain and serve immediately in heated pasta dishes, with the sauce spooned over and sprinkled with the chopped fresh oregano.

500 g/1 lb good-quality dried pappardelle

4 tablespoons torn fresh oregano, to garnish

olive and parsley sauce

125 ml/4 fl oz olive oil

1 small shallot, finely chopped

125 g/4 oz pitted green olives

2 garlic cloves, crushed

4 sun-dried tomatoes, finely chopped

grated rind and juice of 1 lemon

1 bunch of fresh, flat leaf parsley

1 bunch of fresh oregano

salt and freshly ground black pepper

Serves 4

Baked green gnocchi
with creamy onion confit

**A warming dish with sweetly flavoured
onions in a cream sauce, baked with green
spinach gnocchi and Parmesan.**

To make the onion confit, heat the olive oil in a large
pan, then add the sliced onions and stir. Cover the
pan with a lid and cook until softened and golden.
Take care not to burn the onions, and add a little
more olive oil if necessary.
Add the remaining confit ingredients and
gently heat through.
Meanwhile, cook the gnocchi in boiling salted
water for a few minutes until they rise to the surface.
Drain and stir into the sauce.
Spoon the mixture into a gratin dish, sprinkle with
the Parmesan, and cook in a preheated oven at
200°C (400°F) Gas Mark 6 for 15–20 minutes.
Sprinkle with pepper and serve immediately.

500 g/1 lb fresh
gnocchi, spinach
or plain

50 g/2 oz fresh
Parmesan cheese,
shaved or grated

freshly ground
black pepper

onion confit

2 tablespoons olive oil

500 g/1 lb onions,
finely sliced

1 garlic clove, crushed

150 ml/¼ pint
single cream

2 eggs, lightly beaten

Serves 4

Pumpkin ravioli
with a pumpkin and dill sauce

Olive oil and chopped dill are used to flavour
both the sauce and the filling in this
gorgeous, golden ravioli recipe.

To make the filling, heat the oil in a pan, and sauté
the pumpkin cubes very gently for about 20 minutes
until very soft. Add the dill, salt and pepper.
To make the pasta, mix the egg and flour together,
adding enough water to form a stiff but smooth
dough. Knead for 10 minutes until very smooth.
Cover and rest the dough for 40 minutes. Using a
pasta machine, roll out the dough into a long strip
about 12 cm/5 inches wide. Place spoonfuls of filling
in 2 rows, about 2.5 cm/1 inch from the top and
bottom, and the same distance apart, until half the
pasta sheet is covered. Moisten the edges and the
spaces in between. Fold over the other half of the
sheet, then flatten and seal the sections between the
fillings. Cut out the squares with a zigzag ravioli
wheel. Alternatively, use a ravioli cutter or a
pasta machine with the ravioli attachment.
Chill the filled ravioli until ready to use.
To make the sauce, heat the oil in a pan, add the
pumpkin and shallot and sauté until tender. Add the
dill, salt and pepper. Cook the ravioli in a large pan
of boiling water for 4–5 minutes, until they rise to the
surface. Divide between 4 heated plates, spoon over
the sauce, sprinkle with Parmesan and serve.

1 egg

250 g/8 oz plain flour

water (see method)

ravioli filling

2 tablespoons olive oil

500 g/1 lb pumpkin or
butternut squash,
peeled and finely cubed

1 tablespoon
chopped fresh dill

salt and freshly
ground black pepper

**pumpkin and
dill sauce**

2 tablespoons
extra-virgin olive oil

250 g/8 oz pumpkin or
butternut squash,
peeled and finely cubed

1 shallot or small onion,
finely chopped

2 tablespoons
chopped fresh dill

sea salt and freshly
ground black pepper

freshly grated
Parmesan cheese,
to serve

Serves 4

a wonderful, colourful sauce served with ravioli that can be

made either with a cheese filling or this delicious herby stuffing

Rice

Beetroot risotto

For fantastic flavour, roast the beetroot yourself. This recipe is also delicious served with a spoonful of soured cream.

If using fresh beetroot, roast them in a preheated oven at 200°C (400°F) Gas Mark 6 for 30 minutes. To make the risotto, heat the oil in a saucepan, add the onion and fry until softened. Stir in the rice, add a ladle of hot stock, stir and allow the rice to absorb the stock. Stir in the diced beetroot, then add more stock, stir and simmer until absorbed. Repeat until all the stock is used. Taste, adjust the seasoning and serve, topped with chopped sage and Parmesan.

250 g/8 oz cooked beetroot, peeled and diced, or 500 g/1 lb fresh beetroot, unpeeled

2 tablespoons olive oil

1 small onion, finely chopped

350 g/12 oz Arborio risotto rice

350 ml/12 fl oz hot vegetable stock

sea salt and freshly ground black pepper

to serve

chopped fresh sage leaves

shavings of fresh Parmesan cheese

Serves 4

the very essence of tomato – **intense and concentrated** –

topped with full-flavoured basil pesto

Baked tomato risotto
with green pesto

Although there are lots of wonderful flavours for risotto, tomato and pesto are especially good. Whizz up your own pesto when basil is plentiful, using the recipe on page 21. Otherwise, use good-quality ready-made pesto. The dense texture of plum tomatoes is best suited to this recipe – and it's simple to bake them in advance, very slowly, in a low oven, so their liquid evaporates leaving an intense, concentrated flavour.

Place the sliced tomatoes in a roasting tin with the garlic, sprinkle with salt and freshly ground black pepper, then bake in a preheated oven at 160°C (325°F) Gas Mark 3 for 1 hour. Heat the olive oil in a pan, add the onion and lightly sauté until golden. Stir in the rice, then add a ladle of hot stock, stir and allow the rice to absorb the stock. Add more stock, stir, and simmer until absorbed. Repeat until you have used all the stock. Taste and adjust the seasoning. Serve, topped with the baked tomatoes, a spoonful of basil pesto and shavings of fresh Parmesan.

500 g/1 lb plum tomatoes, sliced

1 garlic clove, crushed

2 tablespoons olive oil

1 onion, finely chopped

350 g/12 oz Arborio risotto rice

350 ml/12 fl oz hot vegetable stock

salt and freshly ground black pepper

to serve

4 tablespoons basil pesto (see page 21)

shavings of fresh Parmesan cheese

Serves 4

Vegetables

Mediterranean
vegetable ragoût

Choose your own selection of vegetables, according to what's in season – baby artichokes and broad beans are very good.

Blanch the courgettes in a pan of boiling, salted water. Remove with a slotted spoon and set aside. Add the onions and boil until soft and tender, then remove with a slotted spoon and set aside.
Blanch the mushrooms in boiling water, then drain. Place the marinade ingredients in a large pan, bring to the boil, add the blanched vegetables and simmer for 10 minutes. Cool, transfer to a bowl, cover and refrigerate for a few days for the flavour to develop. Drizzle with extra-virgin olive oil and serve with crusty country bread.

500 g/1 lb
small courgettes

250 g/8 oz
pickling onions

250 g/8 oz
button mushrooms

extra-virgin olive oil,
to serve

marinade

300 ml/1 pint water

juice of 3 lemons

300 ml/½ pint olive oil

2 sprigs of thyme

1 bunch of fresh
flat leaf parsley

4 bay leaves

1 small celery stalk,
with leaves

10 black peppercorns

10 green olives

Serves 4

Pickled aubergines
for bruschetta or pasta sauce

A useful bruschetta topping or pasta sauce,
sprinkled with chopped sun-dried tomatoes.

Place the shallot, bay leaves, salt, vinegar and water
in a pan and bring to the boil. Add the aubergines
and place a plate on top to keep them submerged.
Simmer 10 minutes until soft, then drain.
Pour boiling water over 4 Kilner jars and place in the
oven at 150°C (300°F) Gas Mark 2 for 5 minutes.
Remove, drain, then add the aubergines, garlic,
chillies and rosemary. Cover with oil, leaving
1 cm/½ inch headroom. Seal the jars, set in a
roasting tin filled with water, and place in the oven at
150°C (300°F) Gas Mark 2 for 15 minutes. Cool, then
chill for 2 days before eating. Use within 1 month.

1 shallot

2 fresh bay leaves

2 teaspoons salt

300 ml/½ pint vinegar

600 ml/1 pint water

1 kg/2 lb baby
aubergines (or large
ones cut into chunks)

8 whole garlic cloves

2 fresh red chillies

sprigs of rosemary

olive oil, to cover

Makes 2 jars of
1 litre/1¾ pints

Roasted peppers
marinated in lemon-flavoured oil

A versatile pepper recipe – ideal on its own,
on bruschetta or with pasta.

Grill the peppers on all sides until the skins are
blistered and blackened. Wrap in clingfilm for
3–5 minutes. Slip off the skins, pull out the stems
and seeds, cut in half, and place in a glass jam jar.
Add the lemon peel, then pour over olive oil to cover.
Chill 2–5 days before using and use within 1 week.

8 red or yellow peppers

peeled rind of 1 lemon

about 300 ml/½ pint
virgin olive oil

Makes 4 jars of
300 ml/½ pint

Spiced chickpeas
with aubergines and tomatoes

Olive oil, used as a braising medium, gives the chickpeas loads of flavour. If you're short of time, canned chickpeas make this Middle-Eastern-influenced dish very quick and easy to prepare. Serve with saffron rice and dishes such as pan-fried lamb fillet.

Heat the oil in a heavy-based pan, add the chopped onion and the garlic and sauté gently until softened and transparent.

Stir in the remaining ingredients, bring to the boil and simmer until the tomato sauce has slightly thickened. Serve immediately.

6 tablespoons extra-virgin olive oil

1 onion, finely chopped

1 garlic clove, crushed

½ teaspoon turmeric

small pinch of ground coriander

small pinch of ground cumin

1 aubergine, diced

500 g/1 lb plum tomatoes, skinned, deseeded and finely chopped

2 tablespoons red wine or stock

375 g/12 oz chickpeas, freshly cooked, or canned

sea salt and freshly ground black pepper

Serves 4

a great main dish for vegetarians –

and also **delicious** as a starter

Orange and lemon
polenta cake

Olive oil can be used to replace butter in cakes, and the type of oil you use will create the flavour. If you don't want it too strong, start with a light oil and move on up to a strong Spanish or Greek oil. Polenta gives the cake a lovely, nutty texture and a pretty green-gold colour. This cake is wonderful served plain with coffee, or as a pudding with crème fraîche and a glass of sweet white wine.

Place the orange and lemon in a saucepan with cold water to cover. Bring to the boil and simmer for about 30 minutes until very soft. Drain and reserve the fruit, then cool. Place in a blender or food processor and purée until smooth. Beat the eggs with an electric mixer until pale and fluffy (about 5–10 minutes), then whisk in the sugar. Mix the flour with the bicarbonate of soda and salt, fold into the egg mixture, then mix in the olive oil. Stir in the polenta and the puréed fruit. Pour the batter into 20 cm/8 inch springform pan. Cook in a preheated oven at 180°C (350°F) Gas Mark 4 for about 50 minutes, until a skewer, inserted into the centre of the cake, comes out clean. Remove from the oven, place on a wire rack, sprinkle with icing sugar and, when cool, remove from the pan. Serve plain, or with whipped cream. or crème fraîche.

. . . and som

hing sweet

1 orange

1 lemon

4 eggs

125 g/4 oz caster sugar

200 g/7 oz plain flour

3 teaspoons
bicarbonate of soda

pinch of salt

300 ml/½ pint olive oil

125 g/4 oz polenta

icing sugar, for dusting

4 tablespoons
crème fraîche
(optional), to serve

Serves 8–10

Index